PLUMBING

REAL LIFE GUIDES

Practical guides for practical people

In this increasingly sophisticated world the need for manually skilled people to build our homes, cut our hair, fix our boilers, and make our cars go is greater than ever. As things progress, so the level of training and competence required of our skilled manual workers increases.

In this series of career guides from Trotman, we look in detail at what it takes to train for, get into, and be successful at a wide spectrum of practical careers. *Real Life Guides* aim to inform and inspire young people and adults alike by providing comprehensive yet hard-hitting and often blunt information about what it takes to succeed in these careers.

The other titles in the series are:

Real Life Guide to Catering

Real Life Guide to the Motor Industry

Real Life Guide to Hairdressing

Real Life Guide to Construction

Real Life Guide to Carpentry & Cabinet-Making

trotman

PLUMBING

Mike Hobbs

Real Life Guide to Plumbing
This first edition published in 2004 by Trotman and Company Ltd
2 The Green, Richmond, Surrey TW9 1PL

Editorial and Publishing Team
Author Mike Hobbs
Editorial Mina Patria, Editorial Director; Rachel Lockhart, Commissioning Editor; Anya Wilson, Editor; Erin Milliken, Editorial Assistant
Production Ken Ruskin, Head of Pre-press and Production
Sales and Marketing Deborah Jones, Head of Sales and Marketing
Managing Director Toby Trotman

Designed by XAB

British Library Cataloguing in Publication Data
A catalogue record for this book is available from the British Library

ISBN 0 85660 905 6

Typeset by Photoprint, Torquay
Printed and bound in Great Britain by The Cromwell Press, Trowbridge, Wiltshire

Real Life GUIDES

CONTENTS

ACKNOWLEDGEMENTS 6

ABOUT THE AUTHOR 6

INTRODUCTION 7

1. SUCCESS STORY 11

2. WHAT'S THE STORY? 13

3. TOOLS OF THE TRADE 17

4. POSSIBLE PROBLEM AREAS 23

5. A DAY IN THE LIFE OF A PLUMBER 27

6. MAKING YOUR MIND UP 33

7. CASE STUDY 1 39

8. TRAINING DAY 41

9. CASE STUDY 2 51

10. CAREER OPPORTUNITIES 55

11. RESOURCES 63

Acknowledgements

Many people heled me with useful ideas for the *Real Life Guide to Plumbing* but I would like especially to thank the following:

Ellen Cheeseman
Carol Cannavan
Steve Evans
Rob Wellman
Kathryn Hopkins-Morgan
Kathleen Houston
Dee Carey Pilgrim
Tim Dindjer
Mina Patria
Anya Wilson

Not forgetting my wife, Maureen, and our children, Anna and Jack, for coping with all the ups and downs of preparation and writing.

Mike Hobbs
January 2004

About the author

Mike Hobbs is a freelance writer and journalist. His work has been published in many broadsheet newspapers and magazines, including the *Daily Telegraph*, the *Independent on Sunday* and *Time Out*. He has won awards for his corporate journalism and screenwriting.

Mike has written the **Real Life Guides** to **Plumbing**, **Construction** and the **Motor Industry** and **E-Commerce Uncovered**, all published by Trotman Publishing. He lives in London with his wife and their two children.

Introduction

It's every home owner's worst nightmare: arriving back at your flat or house to find a giant puddle under the main radiator; or discovering your central heating has decided to die in the middle of winter; or worse still flushing the toilet only for the contents of the toilet bowl to back up and empty themselves on your bathroom floor. In each of these examples there is only one logical course of action open to you – flick through your local telephone directory and find an affordable and reliable emergency plumber. However, that may not be as easy as it sounds, for although we all use the facilities a plumber has installed in our homes, schools or offices each and every day of our lives, the truth of the matter is that at present the UK just doesn't have enough trained plumbers to go around.

It's a sorry state of affairs, for the role of the plumber is essential to the well-being of us of all. 'Plumbers are vital to public health,' says Kevin Wellman, the Operations Director of the Institute of Plumbing (IOP). 'We are talking about clean water and sanitation. If you think of some of the problems the Third World has with clean water supplies you realise how lucky we are. Even now, when we go on holiday to some countries abroad we are told to drink only bottled water, whereas in Britain clean water coming straight from the taps, flushable toilets and plumbed-in central heating systems are all part of our daily lives.'

With the skills of the plumber being so essential, it seems strange to discover they are in short supply, but there are various historical and social reasons for the current shortage. At one time plumbing was perceived as a quite lowly and somehow 'dirty'

Plumbers are vital to public health

profession, suitable only for people with limited academic skills, and this meant a lot of very suitable candidates decided to pursue other career routes, many in office jobs. Because the plumber's role was undervalued it meant plumbers' wages were also lower than they should have been. Then a problem arose with the traditional apprenticeship system, whereby an older, fully trained plumber would take on an apprentice and train them on the job.

According to Kevin Wellman many established plumbers became wary of their trainees going self-employed and 'poaching' their clients. 'Also, a lot of people were put off having apprentices because of the red tape involved in sorting out the funding, etc,' he says. 'The upshot of this is that fewer plumbers were coming through the system, and according to the Construction Industry Training Board (CITB) over the next five years there will be a shortfall of 30,000 plumbers across the United Kingdom.' That's a staggering 6000 new plumbers fewer than needed a year.

One of the knock-on effects of all this has been the rise of the 'cowboy' plumber, under-trained, unskilled and charging a fortune for shoddy work. But the public has become very wary of employing just anybody without proper qualifications or references and now demands plumbers who really know what they are doing. This demand means any plumber worth his or her

I know the satisfaction you can get from working with your hands and with other people

salt can be very well paid for their skills. This is why there has never been a better time to decide to get good quality, academically recognised training as a plumber.

The very fact that the nation is crying out for plumbers means you have a very good chance of earning a decent salary. And both the industry and government are working hard to improve the standard of training and its availability. For instance, registrations for City and Guilds (C&G) accredited plumbing courses went up a staggering 75 per cent in 2002 compared to the previous year, while the numbers actually qualifying at C&G Levels 1 to 3 rose 33 per cent.

remember, somewhere in the United Kingdom at this very minute someone's about to phone for a plumber!

While the number of places on traditional apprenticeships may be falling, college places are expanding to fill the demand. The College of North West London had over 2500 applications for just 150 first-year places in 2003, so it has increased first-year places to 230 for 2004.

'It is vital we get more young people coming through,' says Kevin Wellman. 'This is an occupation that can be the pathway which links into so many things and it's a profession you can make the most of for many years to come, from owning your own plumbing and heating business, through designing the plumbing for major public buildings such as libraries or hotels, to working for a water company as a Water Regulator Inspector.' Kevin Wellman should know as he initially trained as a plumber himself. 'It's a great profession,' he says. 'I know the satisfaction you can get from working with your hands and with other people and the buzz of turning around and being able to say "that bathroom suite, I installed that, I made it work." It's fantastic.'

This guide will show you the types of skills and attitudes you will need to make a career in plumbing a success. There are plenty

DID YOU KNOW?

People with qualifications at National Vocational Qualification (NVQ) Level 3 or (or SVQ3 in Scotland) earn on average 25 per cent more each year than those of the same age and experience with no qualifications.

of stories and advice from other people in the profession to let you know what you might be in for. Obviously you won't learn everything from a book, but it will give you a flavour of the different jobs in the industry, its good and bad points, and help you decide whether becoming a plumber is really for you. It will offer information on what courses and qualifications you will need in order to progress and also what the broader opportunities in plumbing are. Whether you want to work by yourself and for yourself, be one of a team, or oversee your own team in a management capacity, plumbing can be a very fulfilling way of earning your living. This book can help you make an informed decision, and remember, somewhere in the United Kingdom at this very minute someone's about to phone for a plumber!

CHRISTOPHER HANLON

Success story

Christopher Hanlon started in the plumbing industry several years ago and has now worked his way up to be Managing Director of his own company in Glasgow, Scotland.

HOW DID YOU GET YOUR FIRST JOB – AND WHAT WAS THE TRAINING LIKE?

I had a very enjoyable apprenticeship with the company I served my time with. It was a company that carried out work similar to the one I now run, in other words within households. I found it both enjoyable and educational from the point of view that we learned various different trades within the plumbing industry.

WHAT DO YOU DO NOW?

I run a Glasgow-based plumbing and heating company that serves the domestic and household repair market.

HOW LONG HAVE YOU BEEN IN BUSINESS FOR YOURSELF?

I started the business over 15 years ago and we've been constantly expanding ever since.

HOW MANY PEOPLE DO YOU EMPLOY?

We now employ between 30 and 35 staff, mainly depending on the time of year. For a number of years we have also been offering a 24-hour emergency service, but recently

A young plumber should be ambitious, have a solid layer of general knowledge but be willing to learn.

we have set up a separate specialist company to service the emergency market.

HAVE YOU ANY ADVICE FOR SCHOOL LEAVERS WISHING TO BECOME PLUMBERS?

In my experience, a young plumber should have a few simple, overriding attributes and qualities. He or she should be ambitious, and want to do well, and have a solid layer of general knowledge but be willing to learn.

To be successful, the apprentice must be enthusiastic and hard-working, dress smartly and always be respectful of others

CAN YOU IDENTIFY ANY QUALITIES THAT MAKE PEOPLE MORE LIKELY TO SUCCEED?

To be successful, the apprentice must be enthusiastic and hard-working, dress smartly and always be respectful of others, which also means being polite and trustworthy, in order to gain the customer's confidence.

2

What's the story?

Many people only ever get to see a plumber at work when they have an emergency, such as a pipe bursting, or when a major plumbed-in fixture, such as a boiler, needs to be replaced. But what, actually, does a plumber do? Basically, plumbing is all about water, pipes and drainage. A plumber installs, repairs and replaces everything from kitchens and bathrooms to radiators, washing machines and dishwashers. The main areas are sanitation systems (such as toilets and sinks), heating systems (these may be fuelled by oil, gas or a form of solid fuel such as coal), outside guttering work, underground pipework, and even weather proofing on roofs. Some plumbers work in private houses, others work on building sites, still others work as maintenance plumbers for large companies. Whether the working environment is residential, commercial or industrial, you'll find that only a qualified plumber has the skills required to install and service the massive variety of plumbing and heating equipment available.

There's already more than enough work for everyone, and with the constant advances in plumbing and heating technology there are bound to be new employment opportunities for plumbers well into the future.

Although different plumbers do different things, the essence of the job remains the same; describing the job is therefore fairly straightforward and clear-cut.

DID YOU KNOW?

In Tudor times plumbers were very highly paid and regarded as extremely valuable members of society That's because it was such a specialist trade. In some ways it still is.

The main skill you'll need is to be able to **cut pipework**, because all water flows in and out of buildings via pipes. You'll have to do this carefully and accurately, ensuring that all pieces and joints fit together.

The other especially useful skill you'll need to acquire in cases of repair is the ability to **find where the fault lies**. This will obviously improve with knowledge and experience.

You're going to have to quickly pick up the art of **positioning household appliances** (such as boilers, sinks, baths and showers) in the best possible places to ensure they function most effectively. Again, experience will help you greatly.

As you're also responsible for seeing that water can flow freely off the roofs of buildings, to avoid it becoming dangerously trapped, you must learn the skill of **shaping materials** to make this happen. This will often mean cutting lead to fix the roofing and carrying out repairs to the guttering.

It's not enough just to be able to perform the basic skills. As you'll be operating on your own in most cases, you have to be able to **plan** your work accordingly, devising a scheduled order of tasks to make sure you complete the job efficiently and effectively.

Different systems have operational quirks that you should learn and master before you take on their maintenance and repair. Since new systems are coming on to the market all the time, you will need to continually **re-educate** yourself to keep your product knowledge up to date.

It will help you greatly if you have the **ability to draw** well – you'll certainly be using this skill a lot in the job, so it's bound to improve. You'll also need to be adept at interpreting the drawings made by others, checking that they are functionally sound and then putting them into effect.

All the work you do will require you to be able to take quick and accurate **measurements** consistently. Carry pencils and paper at all times to mark down the measurements you make.

You'll need to know how to use a full range of **tools and equipment** such as hacksaws and blowlamps, power drills and wrenches, not to mention different kinds of testing and measuring equipment.

As with the interpretation of drawings, you'll find it helps enormously if you're able to **interpret instructions** from others quickly and clearly. Sometimes you'll think that being a mind reader would be pretty useful in the job.

You'll need to be able to **work in all locations** and in all climatic environments. Although the bulk of the work is inside buildings you'll also have to brave the elements and work outdoors sometimes.

The primary purpose of plumbers is to maintain a flow of hot and cold water for each building, so your aim is to keep pipes clear for the swift removal of all waste products and water from toilets, baths, basins and sinks. Efficient drainage depends on you being able to maintain the **underground flow** of water and waste products into the sewerage systems.

Many plumbers also deal with heating systems that are fuelled by gas, oil and solid fuel and other systems such as cookers driven by the same utilities. If you work with gas you must, by law, be **registered with the Council of Registered Gas Installers (CORGI).** We talk about this in more detail in the Resources section of this book.

You'll need to be prepared to work not only on the guttering to ensure rainwater runs away smoothly but also on the roofs themselves to maintain fully effective **weather proofing** using such materials as sheet lead. Here, you'll have to make sure that overflow systems and cold water supply from the mains both continue to function properly.

You'll find that some plumbing companies specialise in domestic house work (there's an example in this guide) or alternatively concentrate on industrial and commercial work in shops and factories, while others may offer a broader range of services. Plumbers can work in quite a few houses each day on different call-out jobs or remain on a larger site for several months at a time. Only on rare occasions will you work at a fixed location (if you're employed to maintain systems at a hospital site, for instance).

Finally, the conditions you work in can vary massively. You could be working in a cramped and boiling hot loft in summertime, or on a freezing cold and exposed roof fixing guttering or frozen pipes in winter.

By now you should have a better idea of just what it is a plumber does. If you think you can handle all the situations and different aspects of work described here then plumbing may well be the kind of flexible job for you. In the next chapter we will explore in more detail if you have the particular skills it takes to be a success in plumbing.

3

Tools of the trade

As a trained plumber you will find yourself using all kinds of specialist tools during the working day, but the most important tools you have are not those in your toolbox, but those you carry around within you. Some people are just not suited to a lifestyle where they are doing different things in different locations and prefer a more structured environment. Still others love the thrill of rising to a challenge, so a job where problem solving is involved is just what the doctor ordered. Listed below are just some of the skills – or tools – you will need to rely on as a plumber. Run through the list and see if you can recognise whether you have these particular strengths and skills. If you don't, they can be learned; but if you already possess them they will make becoming a plumber a much easier task for you.

- It may seem obvious to say so, but it really is true that to succeed as a plumber you have to be **good at working with your hands**. You will be sawing, wrenching, screwing up and unscrewing connecting pipes and joints every day in this job. There are all sorts of exercises you can do to improve your hand–eye coordination, but usually there has to be a certain amount of talent already there for you to work on.
- If you are good at DIY work around the house, or have an ability to carry out handyman and electrical repair work, plumbing may well suit you. If you feel that you could improve the work you do with your hands, try **refining your skills** by doing some odd jobs in preparation.
- **Being conscious of safety aspects** is an absolute must for anyone working in the plumbing industry. Remember, you will be ensuring the efficient delivery of clean drinking water to

people's homes or, if installing a central heating system, will be responsible for the gas and electricity connections, making sure they are correct and secure. Your own safety is also of utmost importance so following health and safety guidelines makes practical common sense.

- When you're learning your trade you will be working under supervision, and often that will mean carrying out jobs as part of a team. However this situation will not last forever – you will soon have to **rely on yourself**. A lot of the work that you will do as a plumber will be on your own so it is absolutely essential that you are able to inspire yourself to get jobs done. It can sometimes be very difficult to push yourself to do a job that is tricky, or smelly, or involves cramped conditions, but you must grit your teeth and do it. There will be nobody else around to do it for you.
- Plumbers will tell you that many of the jobs they do demand **incredible agility**. By definition, many of the water pipes, tanks and boilers will be hidden away and relatively inaccessible and often you will need the skills and flexibility of a contortionist to get at them – natural fitness is therefore important. So, too, are **strength and stamina**. New baths, sinks, heaters and piping can be incredibly heavy and cumbersome to lift, so make sure you're up to the job by looking after yourself physically.
- In addition to physical stamina, it's important that you have the **mental stamina** to be able to cope with working in a pressured environment. The pressure of dealing with exams may be the only direct experience those of you new to a working life will have. Most of you will be able to manage easily and some may even prefer working under the added stimulus of pressure
- Being able to **communicate well** with people is vital. First, you have to have good listening skills to really understand what it is that customers are telling you, because often their descriptions will be either very basic or confusing. You will

have to interpret what they're saying and get to the root cause of the problem. You must also be able to explain very clearly to customers what exactly it is that you're going to do (or have done) and how much it's going to cost. This is extremely important because if a customer is pleased with the job you have done they will be inclined to use you again and to tell all their acquaintances about you. Repeat business and good word of mouth mean a plumber is guaranteed a large and loyal customer base, which is why looking after your clients well is so important.

- Even if you're working on your own you'll still be in **frequent contact with others**. You'll have to demonstrate that you can get on with people at all levels because you're going to be dealing with a constant flow of colleagues, customers and enquiries from potential customers. So, in addition to the fact that you need to be able to communicate clearly, you will also need to be sensitive to the needs of others.
- Because plumbing's a trade where communication has such a clear priority, you'll discover that having a **good sense of humour** is a great advantage. Sometimes, when things have gone wrong (as they invariably do in any walk of life), it's an absolute necessity. Plumbers are well known for their black sense of humour when having to deal with burst pipes, overflowing toilets and broken boilers.
- **Being reliable** and making sure you're always in a position to do what you say you're going to do is an absolute must. Closely linked to this is turning up when you say you will. These days people have busy lives and many will have to take time off work in order to let you into their homes. No one wants to waste a whole morning waiting for a plumber who never arrives. If you are running late, ring your client and give them a new estimated time of arrival. Obviously when you're working for other people you also have an obligation to them to carry out all the work on time and efficiently. If problems do arise (parts not being available, a job taking longer than you

quoted for) you must talk to your client immediately so you can agree on a course of action.

- Strongly linked to reliability are the personal qualities of **trustworthiness and honesty**. You will often be working alone in customers' homes and they will need to feel sure they can leave you to work unsupervised. You must, therefore, be able to inspire their trust. Providing references and becoming a member of an association such as the IOP are both good ways of reassuring clients. Also, don't just have a mobile phone number, but give people an address and landline they can contact you on as well. Again, you'll find it pays dividends in increased business as word spreads on the grapevine.
- Just about every job you'll do as a plumber requires you to **travel quickly** from A to B (unless of course you're working on installing systems on some large new project), so you will have to learn to drive. This is especially true if you work on your own and have to transport all your tools, hardware and equipment with you.
- A lot the work involved in plumbing is painstaking and delicate, demanding **a great deal of patience**. You may be engaged in a particularly tricky aspect of fitting a system, or trying very hard to replace an especially small and inaccessible part. It's not good enough to be almost right so take your time and make sure it's perfect first time.
- You will have to learn to **maintain your cool and be polite** at all times. However unreasonable your client may be, keep your temper and explain everything clearly and logically – to colleagues as well as customers. **Cleanliness** goes hand in hand with politeness. Customers do equate a good appearance with reliability and honesty. Even if you know you are going to get filthy on the job, arrive looking neat and tidy and keep a dust sheet and a change of clothes in your car or van.
- If there is one subject that all plumbers are likely to use a lot, it is **mathematics**. Whether you're working out angles, totting

up sales figures or calculating your wages, good attention to your maths studies at school and college is bound to help in this career. The same is true of **scientific knowledge**. Anyone who aspires to a job in plumbing should have a sound knowledge of the basic principles of chemistry and physics – and if you want to get involved in design and engineering you'll have to take your understanding of theory and practice to a much higher level. You must also have **good reading skills** to read and understand technical instructions.

- Finally, you must have the ability to **adapt** to different and changing circumstances. No two jobs are exactly the same and you have to be willing to learn about new aspects of your work all the time. You may also have to be **flexible** about your hours of work. If you're on a 24-hour emergency call out service, then you have to be ready to carry out a job at any time, which can play havoc with your social life.

4

Possible problem areas

Just as there are certain skills or personal tools that will help you get on in this industry, there are also certain physical conditions that could hold you back, and certain aspects of the job that may not suit you. Think about these seriously if you are considering a job in plumbing.

- If you suffer from **breathing difficulties** or have a condition such as **asthma**, working in plumbing may not be for you. The dust, damp and changes in air temperature that a plumber comes across can exacerbate existing breathing problems. Carry any alleviating sprays or medication with you wherever you go.
- Similarly, if you suffer from **claustrophobia**, you're going to find some of the work very tough indeed. As mentioned already, many boilers, pipes and water tanks are situated in confined spaces with severely limited access – if you don't like the idea of lying under the sink with your nose in the U-bend and a crick in the neck, then you're probably going to be in trouble.
- If you have a family history of **back problems** the constant lifting and carrying of heavy equipment and hardware could be storing up trouble for you in the future. This is a physically strenuous job and even though you will be taught the correct and safe way to lift and carry at college it is best to bear in mind that many people are forced to leave the industry owing to bad backs.
- You may have difficulty if you suffer from **colour blindness**. It is advisable to check with organisations such as SummitSkills

for more information on how it may affect your chances of becoming a plumber.

- Some parts of a plumber's job require you to work outside, maybe on roofs to deal with pipes and guttering, and if you suffer from **vertigo** you should be aware that this might occasionally cause you some distress. If the idea of huddling on the roof of a three-storey building in snow installing a new central heating ventilation unit gives you a panic attack, then you may have to think again about plumbing as a career.
- If you're **no good at organising yourself**, then you're going to have to learn fast. Plumbers are always on the go, usually dashing from one job to the next, so you'll need to have various essential support materials ready and in good working order.
- The flipside of punctuality is **poor timekeeping**, and if you're someone who always seems to be late then it's a problem you're going to have to solve very quickly. This shouldn't be too arduous – you'll just have to learn to aim to arrive ten minutes early in the morning and try to keep that up throughout the day.

By now you should have a pretty good idea about what a job in this industry actually entails and what it takes to be successful in plumbing. If you still believe this could be the career for you, then complete the short quiz below. The answers will enable you to see if you are really suited to working in this industry. Just tick either true or false to each of the following statements.

Do you like to see the results of your work immediately?
TRUE/FALSE

Do you believe that variety is the spice of life?
TRUE/FALSE

Do you think customer service is really important?
TRUE/FALSE

Do you have the patience to take great trouble over little things?
TRUE/FALSE

Do you think teamwork is a vital part of the job?
TRUE/FALSE

Do you find the prospect of working in an office a bit depressing?
TRUE/FALSE

Do you enjoy working with your hands whilst still using your brain?
TRUE/FALSE

If you answered 'TRUE' to all or most of these questions, then you've probably got the right instincts and temperament to be a plumber.

If you've answered 'FALSE' to most of these questions, then you will probably find a career in a different industry is more suited to you. It might be a good idea to check exactly what you want to do and look at other career options.

If your answers are fairly evenly split between 'TRUE' and 'FALSE' you may want to think very carefully about everything that being a plumber means. Read on, but be ready to make some changes to your expectations.

5

A day in the life of a plumber

A busy plumber talks us through his working day and answers other questions about his job.

Steve Evans is a mature trainee, who switched careers reasonably late in life in order to take up an apprenticeship as a plumber. He works mainly in the London area.

HOW DID YOU GET YOUR FIRST JOB – AND WHAT WAS/IS THE TRAINING LIKE?

I got my first job as a follow-on from my first training placement. The placement lasted ten weeks, which gave my prospective employer plenty of time to find out whether I was punctual, diligent and could work at a reasonable speed, and he still offered me a job working with him!

I didn't get paid very much (about £200 per week) but I was lucky to be getting broad experience rather than simply doing central heating, for example. It was difficult for me, aged 36, to find anyone who would train me. I did a lot of research, most of which led to disappointment. However, through LearnDirect I found out about a National Vocational Qualification (NVQ) course at the Community Refurbishment Scheme in Deptford, London. Amazingly, they not only interviewed me but also offered me a place. The tutors were good and very knowledgeable. They teach you how to bend pipes accurately, prepare pipe ends perfectly, ready for soldering, work out the precise fall (downward slope) required for a certain length of basin waste pipe and so on.

Then, when you go on site and are spotted measuring fastidiously or 'wire wooling' a pipe end, you get shouted at for taking too long. Doing the NVQ, I discovered, was very much down to my own commitment to get on and complete the necessary practical and theory work. Several people on the course simply did not do this and ended up not being able to sit the exam. That was sad because nobody was incapable of doing the work.

WHAT HAS HELPED YOU TO PROGRESS?

It sounds stupid, but working with someone who realises you want to do plumbing helps you to progress really rapidly. My boss threw me straight in at the deep end, plumbing in bathrooms and kitchen sinks. Some trainees spend at least a year carting tools and fittings around and generally labouring before they get a look-in on a blowtorch. On placement, make sure early on that everyone knows you're a plumber, not a labourer. That doesn't mean you say no to jobs that don't involve pure plumbing, it just means you want the opportunity to try the real plumbing work.

CAN YOU SUMMARISE A TYPICAL WORKING DAY?

When you get put on a placement you realise that the job is 95 per cent 'heavy grind' labouring: shifting boilers, radiators, baths, toilets. Removing the old stuff, ready for replacement, is often the most work, especially if it involves taking out a hot water cylinder.

On placement, make sure early on that everyone knows you're a plumber

A typical working day (starting at 8am) for me, say on a refurbishment site, might be fitting a bathroom followed by a kitchen sink. This usually involves around 20 trips to the plumbing store (often a big metal container) and skip, which are rarely conveniently sited.

In the morning I start by 'ripping out' the old bathroom: knocking tiles off the wall; isolating the water supply; cutting pipework, removing the old heavy bath, basin and toilet (yes, they are mostly disgusting to handle) and placing them in the skip. This involves a fair amount of water and mess, which then has to be cleaned up. I fetch the new heavy bath and lug it up two flights of stairs, tripping over various obstacles and squeezing through narrow doorways. After this, I 'make it up' with feet, taps, a waste and overflow and, often, bath handles. I then set the bath into position on 2" × 3" timbers, and level it.

After that, I plumb the bath with copper pipe and soldered fittings and plastic solvent weld waste pipe and fittings. Most of this time I am lying on my back on a dirty floor, or kneeling, or bent double. With that finished, I fetch the basin and make that up. I place it on its pedestal and get it to look level (unlikely that a spirit level will help you with this because basin moulds don't seem to be very accurate). With it then screwed to the wall I plumb it in. This is followed by the toilet, which rapidly reveals to me that the floor is very far from level and a plywood plinth must be constructed. In the afternoon, having waited for a kitchen fitter to fit the sink base unit, I do the same to the kitchen sink and hook up the washing machine and dishwasher too. Going-home time is usually around 3.30pm. (*It is important to note that this is Steve's personal experience; the length of a working day will vary greatly according to where you you work.*)

WHAT ARE THE BEST (AND THE WORST) THINGS ABOUT YOUR JOB?
The worst things about the job are the dirt of grubbing around on the floor or even underneath it; the pain of getting soldering flux

into one of the many open wounds on your hands; the aching back and knees. The best things are getting ahead of yourself on a price-work contract and being able to go home regularly at 1.30pm; finishing some pipe work and knowing that (a) it looks tidy and (b) it doesn't leak; not having to take work home with you, and feeling pleasantly physically tired, not mentally distressed.

WAS IT EASY TO SWITCH CAREERS, AND HOW BEST SHOULD OTHERS GO ABOUT IT?

Switching careers to plumbing depends on your age and what you have been doing up to that time. It was not at all easy for me to find an opening on a course, and another way to do this at my

think very carefully about whether you really want to do it. If you do, you should find there are plenty of openings through the local job centre

age would be to save up some capital (to live on) and beg a plumber or plumbing company to let you work for free for two months. If they like you, they'll work something out.

HAVE YOU ANY ADVICE FOR SCHOOL LEAVERS WISHING TO BECOME PLUMBERS?

My advice for school leavers is this: you should think very carefully about whether you really want to do it. If you do, you should find there are plenty of openings through the local job centre. Then put a lot of commitment into it and it'll be plain sailing. It's not easy work or easy money, despite what people think at the moment. However, it can pay quite well and it will offer you a great deal of satisfaction.

CAN YOU IDENTIFY ANY QUALITIES THAT MAKE PEOPLE MORE LIKELY TO SUCCEED?

If you don't mind getting dirty and are not over-fussy about other people's dirt, and if you're physically co-ordinated and mentally capable of rudimentary maths then you'll get on. If you're relatively strong or persuasive enough to get others to help you and if you don't mind the grind of labouring which is, as I said before, most of the work, then you'll succeed. Also, you shouldn't be afraid of trying something new or, at least, something you haven't done before. Even a qualified plumber will wander into unfamiliar territory. If you're sensible, able to read instructions (e.g. for an all-singing, all-dancing jacuzzi bath), and confident of your general abilities to adapt, then you'll go far.

WHAT DO YOU THINK THE FUTURE HOLDS FOR YOURSELF AND THE INDUSTRY?

Well, there'll always be work, that's for sure. Water damage frightens people and most will always want to be rescued by a plumber. It's not complicated, but there are some key considerations, second nature to a plumber, which may elude the DIY-er entirely. Personally, I'd like to see plumbing brought into the 21st century with the spread of solar panels for hot water and heating. This is possible now with new hyper-efficient panels, but people, and often especially plumbers, don't know it.

6

Making your mind up

It's almost crunch time – do you feel you've gathered enough information about plumbing to commit yourself to training for the qualifications that will help you be successful? Just before we examine the ways you can get into the industry here are a few of the most commonly asked questions about training for a job in plumbing and what you, personally, can expect to get out of a job in this industry.

WHAT ACADEMIC QUALIFICATIONS DO I NEED TO GET ON TO A TRAINING COURSE?

Even though there are no specific academic requirements for you to be a plumber, you do obviously have to have a high level of mental alertness and, as places on training schemes are currently at a premium, it's best if you can achieve reasonable exam results (for these purposes Grade A–C GCSE for entry into an Advanced Modern Apprenticeship) in English, Maths, Science, Craft and Design and Technology.

Some employers may ask you to take a selection test to see if you are a suitable candidate for working in the plumbing industry. This is to help you as well as your potential employer because there is no point in putting yourself through the training if you are not cut out for all the aspects of the job.

HOW LONG DO I HAVE TO TRAIN FOR?

There is no set length of training because, on the one hand, you can choose to take the qualifications at whatever speed you wish, and, on the other, it depends on how far up the levels of qualification you wish to go. However, if you're considering trying to pass the NVQ Level 1 or 2, it should take two to three years, and progressing to Level 3 should take a further year to 18 months. Training for such qualifications will be discussed in detail in Chapter 8.

WHAT'S THE PAY LIKE?

Good. As a general guideline, you can expect to be earning an annual salary of £25,000 after you have been employed by a company for a period of two years. Whilst you're qualifying you would obviously be earning less than that. If you're working unsociable hours (such as on 24-hour call out emergency services) you can expect to be paid at a higher rate to compensate for the disruption to your life.

Those of you who progress to becoming self-employed can expect to earn between £30,000 and £40,000. There is a great deal of talk about plumbers earning salaries of around £100,000, but this is highly unusual and depends on other factors such as working anti-social hours and running your own business. The plumbing industry also has a pay-related pension scheme, which means that it offers a level of job security that cannot be found in all industries.

DO I HAVE A NORMAL WORKING WEEK?

Yes, you usually do, varying from eight to ten hours each day at the normal times, depending on the terms of your contract.

However, if you choose to work on emergency 24-hour call out (or if that's the type of work that you get offered) then you will naturally have to be more flexible.

WHAT ABOUT HOLIDAYS?

Holidays are standard, offering you on average four working weeks as holiday each year, as well as all the public holidays available (Christmas, New Year, Easter, Bank Holidays, etc). Many companies will also offer extra days as holiday depending on your length of service, so your entitlement may rise to five weeks, or above, the longer you stay with the same employer. If you work for yourself, one of the great advantages is that you can choose your own holiday arrangements.

CAN I CHANGE CAREERS EASILY?

Yes, you can. As you will have discovered, many of the skills that you learn in the industry are readily transferable, and your knowledge of water might give you a great start if you wanted to build a business installing and maintaining swimming pools and hot tubs. You've also learned flexibility, such as the ability to work on different systems that might enable you to succeed as an electrician. The work that you've done on roofs could translate into finding a job in that part of the construction business – indeed, a great many of your skills would fit you well for a job in that industry.

CAN I WORK ABROAD?

You can work in any country that is a member of the European Union (although obviously being able to speak the relevant foreign language will help). If you want to try this out there is a Young Workers' Exchange Programme (18–28) that will give you work experience or training in the country of your choice, for as little as three weeks or up to 16 months.

There are no such guarantees for other parts of the world, but if you have some solid qualifications behind you they should stand you in good stead.

Promotion prospects are very good indeed within all areas of the industry

WHAT ARE THE PROMOTION PROSPECTS?

Promotion prospects are very good indeed within all areas of the industry. Obviously a great deal, if not all, depends on you. If you work and study hard, complete your training swiftly, develop your expertise and show willingness to do more than is necessary, you'll get on just fine. You'll find a discussion of some typical promotional career paths towards the end of the book in the chapter on Career Opportunities.

DID YOU KNOW?

The actor who plays Phil Mitchell in 'EastEnders', Steve McFadden, used to be an apprentice plumber.

WILL I GET HELP IF I WANT TO BECOME MY OWN BOSS?

Yes, there are several booklets and courses available to help you prepare for the various facets of setting up your own company. We will discuss these later in the Career Opportunities chapter.

WHAT DOES THE PUBLIC THINK OF THE INDUSTRY?

Frankly, the popular perception of plumbing is erratic. Anyone who has had a bad experience or been ripped off by cowboy plumbers will naturally tend to think negatively, whereas those who've had swift, efficient and friendly service will focus on the positive sides of the profession.

There is also a certain amount of misunderstanding about how much plumbers actually earn. Some people seem to think that all plumbers are raking it in, when the sober truth is that those who are earning most are those who are working all hours of the day and night. Attitudes are changing slowly and of course it will partly depend on you (and others like you) to restore and maintain plumbing's good name.

WHAT WILL I GET OUT OF IT APART FROM A CAREER?

Basically, you'll get three main things – a strong feeling of satisfaction and self-worth, a chance to meet a broad cross-section of people, and the peace of mind from knowing that your own household plumbing dilemmas are solved for life.

You'll feel satisfied because you are able to help people by solving a constant flood of problems, each of which is of vital importance for them. To be able to put things right will give you a powerful sense of achievement. Furthermore, you'll invariably be meeting many different kinds of people. As in the rest of life, you probably won't want to be friends with them all, but some you may, and communicating with each and every one of them will be interesting on many different levels.

ELLEN CHEESEMAN

Case study 1

Ellen Cheeseman is one of a small but growing number of female plumbers. She works mainly in the south-east of England.

HOW DID YOU GET YOUR FIRST JOB – AND WHAT WAS THE TRAINING LIKE?

I worked with a qualified plumber whilst training then started doing small jobs for myself. It was really difficult at first, knowing what to charge and remembering everything, but once I'd made a big mistake and had to call the same friend in to help me out, everything was fine after that!

WHAT HAS HELPED YOU TO PROGRESS?

Having the determination to succeed. This may sound corny but it's true, because I don't like to be beaten. I also got involved with my local branch of the Institute of Plumbing, which I found to be very beneficial.

CAN YOU SUMMARISE A TYPICAL WORKING DAY?

No, every day is different. That's what I like most about being a plumber!

WHAT ARE THE BEST (AND THE WORST) THINGS ABOUT YOUR JOB?

I think the best part is working for myself and seeing the look on satisfied customers' faces once a job is complete. I can also work the hours I choose so that I am around for my son. The worst bits are all the hours you work

Every day is different. That's what I like most about being a plumber!

for no pay, behind the scenes. It takes ages to keep up with all the paperwork.

WAS IT EASY TO SWITCH CAREERS?

No, not particularly, but if you really want to and can get someone to give you a push, it's possible. I had a lot of support from my family, which helped. Also I wasn't in the position of having to work to contribute to the household.

HAVE YOU ANY ADVICE FOR SCHOOL LEAVERS WISHING TO BECOME PLUMBERS?

Aim as high as you can. Work hard to get all your qualifications and behave like a professional right from the beginning. If possible, try to be multi-skilled by taking on an extra course in plastering or decorating. This will give you many advantages in the future, since your skills will always be in demand. Go for it, it's worth it in the end.

Aim as high as you can. Work hard to get all your qualifications and behave like a professional right from the beginning

CAN YOU IDENTIFY ANY QUALITIES THAT MAKE PEOPLE MORE LIKELY TO SUCCEED?

The two most useful are determination and courage. Also you must have a belief in what you're doing and in yourself.

WHAT DO YOU THINK THE FUTURE HOLDS FOR YOURSELF/THE INDUSTRY?

Who knows – I'd like to build up a large company providing all the services possible to do with plumbing, employing an equal number of women and men plumbers. There are also opportunities to go into teaching plumbing courses. At the end of the day, I believe there will always be work for a professional plumber.

8

Training day

By now you should have a good idea whether or not plumbing is for you, but before you can go out there and start to work you will have to undergo training.

Although there is no upper age limit for training, and there are now many older people who have decided to give up their initial career and retrain as plumbers (apparently the average age of the Institute of Plumbing's 11,000 trainee members is 35), in order to get funding for most courses you must be under 25 years old. That is why going straight from school into a course or an apprenticeship makes good financial sense. (If you are older you may be eligible for a Career Development Loan – more of which later.)

NATIONAL VOCATIONAL QUALIFICATIONS

The main plumbing qualifications are the National Vocational Qualifications (NVQs, or Scottish Vocational Qualifications – SVQs – in Scotland). Most colleges will prefer you to have Maths and Science GCSE grades A–C to take NVQs. NVQ/SVQ Level 2 gives you a basic foundation in plumbing, but the Institute of Plumbing recommends all plumbers to attain Level 3, which is more comprehensive and deals with domestic, commercial and industrial plumbing. The IOP is also developing a Master Plumber Certificate, for those who have already attained Level 3. NVQs/SVQs allow you to learn practical skills on the job whilst training at college on a block or day release basis.

In all there are five levels of NVQ, with Level 1 being the most basic and Level 5 the most advanced. This means that the skill levels in NVQ/SVQ terms can be broadly defined as:

NVQ/SVQ Level 1 – Basic skills, with an introduction to your chosen topic

NVQ/SVQ Level 2 – Completion of Foundation skills

NVQ/SVQ Level 3 – Advanced skills, to give you specialist knowledge

NVQ/SVQ Level 4 – Supervisory skills, or very advanced technical skills

NVQ/SVQ Level 5 – Managerial skills for those who want to progress further.

MODERN APPRENTICESHIP

If you are aged between 16 and 24 you are eligible to do a Modern Apprenticeship (MA, or Skillseekers in Scotland). Foundation Modern Apprenticeships (FMAs) and Advanced Modern Apprenticeships (AMAs) allow you to earn as you learn. You must have at least four GCSEs at grade C to apply. The AMA leads to the equivalent of NVQ Level 3. In both cases you have to do some on-the-job training with a company and some theoretical training at a local college.

You can do this at the same time or in stages. So you'll take one of three routes:

1. Attend a college as a part-time student

You usually go either one day a week or in weekly blocks. You can try to pass the theoretical side of the qualification as fast as possible to help you move on to the practical stage of working with a company.

2. Join a training programme

If you can, it's best to do the practical and theoretical stages at the same time because it helps you to relate all the pieces of learning to each other. Training programmes will be run either by an employer or by an organisation that works with employers to provide training such as SummitSkills (see Resources chapter).

You will be helped to build up a National or Scottish Vocational Qualification in all the necessary subjects at the right NVQ/SVQ levels for your chosen career (usually a combination of Levels 1, 2 and perhaps 3).

3. Go straight into employment

If you cannot find a position on a training course that suits you, it might be possible to join a company, establish yourself in the job, and then start your formal training later.

Once you have qualified as a plumber you can continue training to reach NVQ/SVQ Levels 4 and 5 if you wish. You can also take qualifications in related subjects such as welding and electrical installation (and all the other skills related to Building Services Engineering) that have appropriate NVQ/SVQ levels.

Note that there are changes to the qualification system currently taking place, including:

- the reorganisation of the Edexcel (BTEC) national and higher national system;
- the introduction of national qualifications in Scotland;
- the introduction outside Scotland of Advanced Subsidiary (AS) levels, vocational A levels, three unit and six unit GNVQs and the Key Skills Qualification.

WHAT TRAINING COURSES EXIST?

Training courses, as stated earlier, are a mixture of on-the-job training from the company you are working for and off-the-job, theoretical training from a local college. There are many of these colleges in all parts of the UK.

You can find lists of possible companies and colleges on the Institute of Plumbing website. There are also full details of colleges and training centres on the SummitSkills site and the British Plumbing Employers' Council (BPEC) website. All these addresses and lots more useful contact information are contained within the Resources chapter of this book. Alternatively, scour your local phone book for possibilities. It's probably a good idea to keep a record of the progress of all your enquiries.

There are five main ways of being taken on for training:

- you are nominated by national training providers;
- you are nominated by local training providers;
- you are nominated by careers services/Connexions;
- you are nominated by your local employment agency;
- you are nominated by your employer.

All routes require you to take some action on your own behalf.

National Training Providers

Each country in the UK has a designated national training provider for the profession. SummitSkills is responsible for skills development throughout the UK and, in conjunction with various national bodies, forms these plans.

Local Training Providers

All the countries are subdivided into smaller areas, where the advisers will obviously know in greater detail what is happening in your locality. Look at the Learning and Skills Councils for England, Scottish Enterprise and their Local Enterprise Councils for Scotland and, in Wales, at Education and Learning Wales.

Careers Services

Both your local and your school careers services can refer you for training in the profession if you can demonstrate that you have the potential to succeed.

Connexions

Connexions is the government's support service for all young people aged between 13 and 19 in England. It can give you advice and help on starting your career and offer you personal development opportunities.

Local Employment Agency

Your Job Centre or local employment agency can refer you for training in the profession. Again, you must show you have the potential to succeed.

Employers

You can take the initiative and, if you are already in work, convince your employer to apply for you to receive training. Alternatively, you can approach prospective employers and undertake to carry out the training as a condition of employment.

THE NEW DEAL

The New Deal applies particularly in an area where job opportunities are limited. The scheme supports you if you're having difficulty in finding work by helping you get training towards a particular career.

If you decide to participate in the scheme you will have the support of a New Deal adviser who will help you to find a job and get the appropriate training. This training will normally lead to a part- or full-time NVQ.

The New Deal scheme is administered through the Employment Service. To find out the contact details of the service nearest to you, it's best to refer to your telephone directory or your local careers office.

ADVANCED MODERN APPRENTICESHIPS IN SUPERVISION (LEVEL 3) AND DESIGN (LEVEL 4)

SummitSkills has recently brought in its AMA Technician Framework in England and Wales for NVQ Level 3 covering Supervision for Project Engineers, and NVQ Level 4 covering Design for Design Engineers.

These are forerunners of the new Building Services Engineering NVQs, the framework of which will be in place by Autumn 2004.

Because SummitSkills covers the electrotechnical, heating, ventilating, air conditioning and refrigeration industries as well as plumbing, the guidelines for this new framework are necessarily broad.

MATURE TRAINEES

Because so many Mature Trainees (generally defined as anyone over 24) have been expressing interest in trying to switch career to become plumbers, there is now an accepted pathway for you to follow.

As a Mature Trainee, you can expect to achieve the same qualification as young people proceeding through a Modern Apprenticeship framework, which leads to achievement at NVQ Level 3.

This would include passing the knowledge content of the Certificates in Plumbing, picking up experience in the workplace and ensuring that you undertake the appropriate upgrading of Key Skills in communication, application of number and information technology to match the requirements of a plumber.

You'll be pleased to hear common sense prevails here. If you are considerably older or have been involved in a related profession (say, construction work) then you'll find the time you need to finish training will be reduced. This will obviously depend on the match of the range of existing skills – the more you've learned already, the shorter your training will be.

However, most of you will not be able to find full-time college places. Entry for many of you will be achieved by proceeding straight to employed status. This would mean your working with an employer and attending a centre (normally a college) on a part-time basis.

Naturally you will be given targets to attain. You will have to pass the two stages of certification – the Certificate in Plumbing (Intermediate) and the Certificate in Plumbing (Advanced). Your employer will provide the work-based part of the programme and the training route will normally be managed by a Learning Provider, who will receive the training funding from the local Learning and Skills Council (or from a Devolved Regional Authority in Scotland and Wales).

OTHER METHODS OF LEARNING

Some employers may have their own in-house training programmes that conform to the standards set by the training authorities and give you the equivalent of the NVQ/SVQ qualification.

Distance learning packages are also available, which give you all the academic content you need to complete that side of the qualification, although you will need to buttress it with some practical training to gain your NVQ/SVQ.

BPEC/PHIA APTITUDE TEST

This special test has been developed by the Plumbing and Heating Industry Alliance (PHIA) in conjunction with BPEC and is available as part of a free CD-Rom from SummitSkills. It will give you an indication of whether you might have the skills and qualities to succeed as a plumber.

When you have completed your aptitude test on the CD-Rom, record your score and the consequent result, and include a copy of the information with your application to any potential employers, not forgetting to keep a master copy for your own records.

DID YOU KNOW?

The average pay for a plumber can rise depending on the project he or she is working on. For instance, workers on the massive Heathrow Terminal Five project are earning up to £50,000.

If you are not successful on the practice aptitude test, seek advice from your local college on the best ways to upgrade your skills. Remember you can always prepare and practise so that you can take the test again.

FUNDING

You'll have noticed some references to funding earlier and the good news is that there is funding available to support your training and assessment, provided either through the local Learning and Skills Council (or Devolved Regional Authority if you live in Wales, Scotland or Northern Ireland), or from Job Centres if you are unemployed.

Once you've applied, the availability and amount of funding, the timescale over which it is provided, and when it will be paid, will all be identified as a result of your initial assessment by the Learning Provider. As you'd expect, some regional variations tend to occur but, whatever happens, you or your employer will be expected to make a contribution to this process. In other words, the funding will not completely cover your costs.

The Department for Education and Skills (DfES) offers a Career Development Loan for those undertaking a two-year vocational training course. It is available to employed, self-employed and unemployed people and covers full-time, part-time and distance learning courses. Basically, the DfES pays the interest on the loan for the length of the course and up to one month after it finishes. Details are given in the Resources chapter.

The following guide sums up the various routes into a career in plumbing, from leaving school right to the top of the profession.

access to

PLUMBING

NO QUALIFICATIONS

ENTRY LEVEL QUALIFICATION

FOUR GCSEs (A–D) grades 1–3
GNVQ/GSNVQ level 1
selection interview

ON THE JOB TRAINING

APPRENTICESHIP ✦ TRAINEE SCHEMES

ADVANCED MODERN APPRENTICESHIP (England) **SKILLSEEKERS** (Scotland) **MODERN APPRENTICESHIP** (NI) **MODERN APPRENTICESHIP** (Wales)	e.g. **BPEC** (British Plumbing Employer Council) **CITB** (Construction Industry Training Board)

e.g.
ASSISTANT PLUMBER
PLUMBER

CREDITS/FURTHER LEARNING

ON THE JOB QUALIFICATIONS ✦ PROFESSIONAL BODIES

NVQ/SVQ level 1 **BTEC HNC/HND** **Full-time/part-time/distance learning**	e.g. **CORGI** (Council for Registered Gas Installers)

CAREER OPPORTUNITIES

DEVELOPMENT OPTIONS

HIGHER EDUCATION ✦ MANAGEMENT ✦ FREELANCE

9

PAUL TUDOR

Case study 2

Paul Tudor is a 37-year-old plumber and ACS and CORGI accredited gas engineer. He is currently self-employed and works in the Kingston and Surrey area.

'I started out as an apprentice with British Gas back in 1982 and I worked for them for seven years before switching to TRANSCO, their safety division and basically I worked for them right up until four months ago when I went self-employed.

'At school I didn't know what I wanted to do so the careers adviser gave me some brochures on plumbing and the fire service. Well, the fire service wasn't recruiting but British Gas was and so I wrote off to them. At the time there were 85 people going for just two jobs and I was lucky enough to get one of them.

'I was a typical kid who had never had any interest in plumbing but this was a great opportunity and I actually settled into the role fairly comfortably. There were a lot of other lads working with me who shared the same sense of humour and so I did feel at home. I started by doing domestic appliances with British Gas and I liked the fact you were never tied down to the same place any day because I just knew I didn't want an office job. I worked fairly regular

The best things about my job are the money and the sense of satisfaction you get when you do a job well hours until I transferred to TRANSCO and of course they have 24 hour call out because of emergencies. As an apprentice you work with another engineer, then you go out on basic duties for a probation period before being sent out on your own. Generally, when you first start with British Gas you are servicing boilers and when you start with TRANSCO you are resetting meters. As you progress you go on to boiler breakdowns and emergency stuff.

'I think the training has made me so aware of the dangers of gas I was never nervous about working with it. It means you never go into situations where you don't know what you are doing. I suppose if you sat back and thought about it you might get nervous. I have had some hairy moments though. I remember one instance when I was working for TRANSCO when I was called out at 11pm to a block of flats. The smell of gas hit me as I walked into the block and when I walked into the lift I immediately got explosive readings on my detector where the gas had settled in the lift shaft. It turned out someone had left their gas cooker on full blast and everyone in the block could smell it except for them! British Gas and TRANSCO have to keep up with the latest health and safety standards and initiatives and that meant throughout my career with them I was sent on lots of refresher courses and training courses to keep in touch with all the new technology.

'I became self-employed mainly for family concerns and now I am working on a big project on a block of converted flats and I am subcontracted back to British Gas for three days a week. At the flats I am installing all the boilers and doing all the plumbing including showers, baths, sinks, toilets and absolutely all the pipe work. It's a massive job but I really like it because there is a big group of us working together and the banter is great. Also, the electrician is my brother-in-law and the plasterer is my father-in-law so we really are one big family.

'The best things about my job are the money and the sense of satisfaction you get when you do a job well. Say someone has gone out and bought a nice decorative gas fire and I put it in and it's a neat job with no pipes showing and I turn it on and leave them with it glowing up nicely. That's a lovely feeling to leave someone happy and cosy, that's really fulfilling.

'The worst thing is most of your life is spent on your knees. As you get older you really start to feel it and it really does take its toll because if you are doing pipe work you will generally be on your knees. Also, it's fiddly, intricate work and you really have to concentrate so it is mentally hard. Just think about the bottom of a boiler – you are going to have to connect a gas pipe, a vent pipe, a hot water pipe, a cold water pipe, and pipes to all the radiators and they all go in different directions so there is a lot of thinking involved. Also, people dying because of something I have or have not done is my greatest fear and so I try to do the job to a level of safety above and beyond the call of duty.

'Now I just want to make a decent living at what I do but if the opportunity of a big contract came up then I would start employing people. I'm finding I'm getting a lot of work through word of mouth and I think I get recommended because I always try to be neat and tidy, I'm safe and I'm legal and I always give people a card with my address on it so they know where to find me if there are any problems.'

10

Career opportunities

Of course it's entirely possible that you may be more than happy to stay as a basic 'wet only' (i.e. no gas work) plumber for the whole of your working life. Some people are also quite content to stay with the same employer for most or all of their working days. However the possibilities and opportunities to advance up the plumbing career ladder are enormous and some of you may wish to move on and develop your career into broader fields. How far you take it will depend on how much time and effort you are willing to put into more advanced training and education.

With more experience you could **start your own business.** Because it is relatively simple to do, many plumbers who work in the domestic sector start up their own business, giving them the chance to earn greater rewards (but with greater risks and the likelihood of far longer hours). The lure of being your own boss is always attractive. If you prove successful, one day you could be taking on trainees or apprentices of your own. However, you must be prepared for all the extra paperwork that managing your own company will undoubtedly bring with it, not to mention the special problems involved in employing other people if you do choose to expand. The best thing to do is to ask someone you know who has set up their own business, take their advice about the potential benefits and pitfalls, and make your own decision about whether you want to make this one of your aims for the future.

There are also plenty of career openings for plumbers to work as **lecturers** and train other plumbers in various colleges of further

This is an industry with real opportunities for advancement. As you go through the training and discover where your strengths lie you will be able to map out a future career path. The diagram below shows options that will open up to you once you have trained.

CAREER OPPORTUNITIES

TRAINING IN KEY PLUMBING SKILLS

CUTTING, BENDING AND JOINING OF ALL PIPE WORK
CUTTING, SHAPING AND FIXING OF SHEET ROOFING MATERIALS
DIAGNOSIS OF FAULTS IN HEATING SYSTEMS

MORE EXPERIENCE NVQ LEVEL 3 UPWARDS

FINISH APPRENTICESHIPS IN ALL PLUMBING SKILLS AREAS

FURTHER EXPERIENCE NVQ LEVEL 3/4 TRAINING FOR MANAGEMENT

SUPERVISOR/TEAM LEADER

FURTHER EXPERIENCE NVQ LEVEL 4 & 5

FOREMAN/SITE MANAGER ✦ PLANNING
MANAGEMENT ✦ TEACHING

education and training centres. The only danger here is that you might start to get a little detached from all the new advances that are hitting the industry. One way around this is to combine your lecturing with continuing to work as a jobbing plumber. Another is to attend the various technical evenings and seminars run by the Institute of Plumbing to keep its members up to date with developments. In fact, becoming a member of the IOP is a very good way to further your career. Again, listen to what lecturers you know have to say, and see whether this step might be right for you.

If you're working for one of the medium to large businesses, they will often present you with the opportunity for advancement into **technical, supervisory** or **managerial** jobs. This may include the supervision of installation work, business management, maybe the design of complex systems or providing project estimates. If you are clear that this is what you'd prefer to be doing rather than just straightforward working with the tools, then by all means explore the option. The main point here is that you will probably have the chance to watch your own supervisors or managers at work so you can see for yourself what sort of problems they have to deal with in their working lives. If the thought of following them inspires you, you've given yourself a good start towards achieving this goal.

DID YOU KNOW?

Legendary football international Tom Finney, who scored 30 goals for England and won 76 caps between 1946 and 1958, was also setting up his business as a plumber at the time. In the days before huge wages and marketing endorsements, Tom often used to fit in a morning's work before turning up to play for his (only) club – Preston North End. His business is still flourishing today.

As was mentioned earlier on in this book, there should be no shortage of jobs on the market if you are really determined to find one. However there is no guarantee that the vacancies will be evenly distributed throughout the country so it's best to check with your local careers office before you start applying.

Careers advisers can put you in touch with plumbing companies in your area, or you can look them up in the Yellow Pages and write to them yourself. You can also use Internet search engines to look for plumbers or heating engineers.

Alternatively, if you want to find approved qualified plumbers, you can either go to the Institute of Plumbing website or check out your local trade directory. Remember to keep a note of the progress of your enquiries. It's best to be thorough and methodical in your search. For instance, it might be a good idea to start by contacting prospects in your local town or area. If that draws a blank, then spread the net to bring in adjacent towns and areas – and go on broadening your approach until you get a favourable response.

The important point is that you must not get downcast by people turning you away. It's a highly competitive job market and employers get many applications for each position so you must not take any rejections personally. Eventually your persistence will bear fruit.

PREPARING YOUR CV

It's a good idea to review constantly the way that your message is coming across to possible employers. Your character and personality cannot register with them until they've seen you. Bear in mind you've got to convince people to grant you an interview in the first place. Here are a few suggestions on presenting yourself on paper to potential employers by writing an eye-catching CV and a punchy covering letter.

Remember what the function of a CV is, namely to make you stand out from the crowd and get you an interview. So keep your

CV short, no more than two pages, but provide employers with the sort of detail that will bring you to their notice.

If you haven't yet had many, or even any jobs, don't worry. The sort of details that it will be useful to concentrate on are your achievements and the skills you've acquired – and make sure that those skills correspond with those that any reputable company will be looking for.

Lay out your CV neatly and clearly, whether it is being sent by post or email. Check carefully to make sure everything reads well and is free from mistakes. Compose your covering letter with great care so that it conveys your enthusiasm for joining the company without indulging in overselling. For a full examination of how to present yourself on paper, try *Winning CVs for First-time Job Hunters* by Kathleen Houston (Trotman), which has excellent advice on every stage of the process.

Here is a list of the most important things you can do to make sure you are the candidate a prospective employer will be willing to take on:

- Do your research on your target company – in depth.
- Prepare a relevant, customised CV.
- Show you've got the energy, commitment and aptitude to be a plumber.
- Gain any work experience you can.
- Read everything about the industry you can get your hands on.
- Apply for training programmes at your natural level (e.g. Modern Apprenticeships).
- Snap up any free courses.
- Develop other related key skills.
- Be ready to do anything to gain a start.
- Take all the luck you can find (but remember the harder you work, the luckier you get).

FUTURE INITIATIVES

There are plans to introduce three changes that will have a significant impact on all of you who wish to train as apprentices:

- free training to qualify workers to Level 2;
- Modern Apprentice age cap lifted to allow all ages 25+ to train under the scheme;
- support for Level 3 training to enhance technician and higher craft skills.

These changes underline the commitment the government and training bodies are making to ensure our modern-day plumbing workforce is as professional and committed as it can be.

If you have made it this far through the book then you should know if **Plumbing** really is the career for you. But, before contacting the professional bodies listed in the next chapter, here's a final, fun checklist to show if you have chosen wisely.

THE LAST WORD

✔ TICK YES OR NO

DO YOU LIKE WORKING WITH YOUR HANDS? ☐ YES ☐ NO

DO YOU LIKE WORKING WITH PEOPLE? ☐ YES ☐ NO

DO YOU WANT A JOB WHERE YOU WILL BE DOING SOMETHING DIFFERENT EVERY DAY? ☐ YES ☐ NO

ARE YOU SELF MOTIVATED AND ABLE TO THINK ON YOUR FEET? ☐ YES ☐ NO

ARE YOU ABLE TO COMMUNICATE EFFECTIVELY WITH LOTS OF DIFFERENT PEOPLE? ☐ YES ☐ NO

ARE YOU A SELF STARTER, ABLE TO TAKE CONTROL AND RESPONSIBILITY? ☐ YES ☐ NO

If you answered 'YES' to all these questions then
CONGRATULATIONS! YOU'VE CHOSEN THE RIGHT CAREER!

If you answered 'NO' to any of these questions then this may not be the career for you.
However, there are still some options open to you,
for example, you could work as a Sales person at a Plumbers' Merchant or DIY store.

11

Resources

In this section you will find all the addresses, telephone numbers and websites for the relevant government and industry advisory and training bodies for the plumbing industry. There is also a list of publications you may find useful to read.

TRAINING AND ADVICE

SummitSkills
Gear House
Saltmeadows Road
Gateshead
Tyne and Wear NE8 3AH
0191 490 3306
www.summitskills.org.uk

This is the Sector Skills Council for the electrotechnical, heating, ventilating, air conditioning, refrigeration and plumbing industries. It aims to champion the training and development of all these industries.

British Plumbing Employers' Council (Training) Ltd (BPEC)
14 Ensign House
Ensign Business Centre
Westwood Way
Coventry CV4 8JA
024 7647 0626
www.bpec.org.uk

BPEC operates as a charitable trust funding appropriate training-related initiatives for the plumbing industry. Its website gives some great information on training opportunities.

The Institute of Plumbing (IOP)
64 Station Lane
Hornchurch
Essex RM12 6NB
Tel: 01708 472791
www.plumbers.org.uk

Founded in 1906, this is the UK's professional body for plumbers and others in the industry and has over 11,000 members. A registered educational charity, it is also secretariat to the World Plumbing Council. The website has a very good careers section and links to other related bodies as well as a very interesting section on women in plumbing.

Institute of Domestic Heating Engineers (IDHE)
Dorchester House
Wimblestraw Road
Berinsfield
Wallingford OX10 7LZ
Tel: 01865 343096
www.idhe.org.uk

IDHE is an independent, non-profit-making professional body founded in 1964. It is solely for the domestic heating industry.

Scotland and Northern Ireland Plumbing Employers' Federation (SNIPEF)
2 Walker Street
Edinburgh EH3 7LB
0131 225 2255
www.snipef.org.uk

Founded in 1923, SNIPEF is the trade association for businesses involved in installing and maintaining plumbing and heating systems throughout Scotland and Northern Ireland.

Plumbing and Heating Industry Alliance (PHIA)
www.phia.org.uk

Born out of the Plumbing Forum the PHIA has 22 members, including the Worshipful Company of Plumbers, the Institute of Plumbers, and the Institute of Domestic Heating Engineers. The website contains contact details for each of its members.

The Council for Registered Gas Installers (CORGI)
www.corgi-gas.com

This is the national watchdog for gas safety in the UK and registration is a legal requirement for both businesses and self-employed people working on gas fittings, so if you want to work in this area once you are qualified you will have to register. Registration details can be found on the website.

Construction Industry Training Board (CITB)
Bircham Newton
King's Lynn
Norfolk PE31 6RH
01485 577577
www.citb.co.uk

The CITB is committed to training professionals to a very high level. It has some very good pamphlets and brochures offering more information; alternatively have a look at the website.

City and Guilds
1 Giltspur Street
London EC1A 9DD
020 7294 2468
www.city-and-guilds.co.uk

City and Guilds is the leading provider of vocational qualifications in the UK. It has five different levels of qualification, with 1 based on the lowest competence level and 5 based on the highest, and it offers everything from NVQ and SVQ to Modern Apprenticeships and Higher Level Qualifications. The website lists all the qualifications it provides in plumbing.

Learning and Skills Council
Modern Apprenticeship helpline
08000 150600
www.lsc.gov.uk
www.realworkrealpay.co.uk

Launched in 2001, Learning and Skills Council now has 48 branches across the country. It is responsible for the largest investment in post-16 education and training in England and this includes further education colleges, work-based training and workforce developments. Its realworkrealpay website is specifically aimed at those who would like to do Modern Apprenticeships.

For MAs in Scotland:
www.modernapprenticeships.com or
www.careers.scotland.org.uk

In Wales:
www.beskilled.net

Department for Education and Skills (DfES)
Packs available from 0800 585505
www.dfes.gov.uk

If you are undertaking a vocational training course lasting up to two years (with one year's practical work experience if it is part of the course) you may be eligible for a Career Development Loan. These are available for full-time, part-time and distance learning courses and applicants can be employed, self-employed or unemployed. The DfES pays the interest on the loan for the length of the course and up to one month afterwards. You can also find out which colleges are Centres of Vocational Excellence for plumbing at the website www.dfes.gov.uk/coves

Connexions
www.connexions.org.com
www.connexionscard.com

The Connexions service has been set up especially for young people aged 13 to 19 and offers advice, support and practical help on many subjects including your future career options. In the Career Zone on the site you will find Career Bank, offering information on training and jobs in plumbing.

Edexcel
Stuart House
32 Russell Square
London WC1B 5DN
0870 240 9800
www.edexcel.org.uk

Edexcel has taken over from BTEC in offering BTEC qualifications including BTEC First Diplomas, BTEC National Diplomas and BTEC Higher Nationals (HND and HNC). It also offers NVQ qualifications. The website includes qualification 'quick links' and you can search by the qualification or the career you are interested in. Edexcel is currently being reorganised and all course and qualification information should be checked with them.

New Deal
www.newdeal.co.uk

If you are an older individual looking to change careers and you have been unemployed for six months or more (or receiving

Jobseekers Allowance), you may be able to gain access to NVQ/SVQ courses through the New Deal programme. People with disabilities, ex-offenders and lone parents are eligible before reaching six months of unemployment. Check out the website for more information.

Qualifications and Curriculum Authority (QCA)
83 Piccadilly
London W1J 8QA
020 7509 5555
www.qca.org.uk

In Scotland:
Scottish Qualifications Authority (SQA)
Hanover House
24 Douglas Street
Glasgow G2 7NQ
Customer Contact Centre: 0141 242 2214
www.sqa.org.uk

These official awarding bodies will be able to tell you whether the course you choose leads to a nationally approved qualification such as an NVQ or SVQ.

PUBLICATIONS

Heating and Plumbing Monthly
Becket House
Vestry Road
Sevenoaks TN14 5EJ
01732 748000
www.unity-media.com

A monthly magazine aimed at heating engineers and plumbers working in the domestic and light commercial sectors.

Plumb Heat
2 Walker Street
Edinburgh EH3 7LB
0131 225 2255
www.snipef.org

A quarterly journal containing information for the Scottish and Northern Ireland Plumbing Employers' Federation.

Plumbing and Heating in Northern Ireland
139–140 Thomas Street
Portadown
Craigavon BT62 3BE
028 3839 2000
Magazines@mainstreampublishing.co.uk

This quarterly contains industry and product news for the Northern Irish market.

Plumbing Magazine
64 Station Lane
Hornchurch RM12 6NB
01708 463114
www.plumber.org.uk

This is the bi-monthly journal of the Institute of Plumbing giving industry news aimed at members.